CW00751356

The Guide to Rocket Science

Peter Berlin

Published by Oval Books
5 St John's Buildings
Canterbury Crescent
London SW9 7QH
United Kingdom

Telephone: +44 (0)20 7733 8585
Fax: +44 (0)20 7733 8544
E-mail: info@ovalbooks.com
Web site: www.ovalbooks.com

Copyright © Oval Projects 2008
All rights reserved, including the right
of reproduction in whole or in part
in any form.

Series Editor – Anne Tauté

Cover design – Jim Wire & Vicki Towers
Printer – J H Haynes & Co Ltd, Sparkford
Production – Oval Projects Ltd.

The Bluffer's Guides ® is a
Registered Trademark.

The Bluffer's Guides series is based
on an original idea by Peter Wolfe.

ISBN-13: 978-1-906042-11-0
ISBN-10: 1-906042-11-X

CONTENTS

Introduction 1

The basics 3
- Weightlessness and gravity 6
- Orbits, trajectories and other encounters 8
- Rocket science in the news 10
- From flaming hell to starlit heaven 11
- Talking through your ears 14

The great pioneers 17
- Wernher Magnus Maximilian Freiherr von Braun 19
- Sergei Pavlovich Korolev 23
- Edward Mukaka Nkoloso 29
- The moral dimension 31

Wild rockets 32
- Otrag 32
- Rotary rocket 34
- The space shuttle 36
- The space scuttle 37
- The toilet question 37

Clever satellites 39
- Global Positioning System (GPS) 39
- Hubble Space Telescope 40
- Iridium 41

Exotic launch sites 43
Be as close to the equator as possible 44
Kourou 46
Cape Canaveral and Vandenberg 47
Baikonur 47
Truth and consequences 48

Movers and shakers in space 51
The customer and the contractor 51
Who is who in the space business 53
The Coca-colonisation of space 56
The human factor 58

To be or not to be a rocket scientist 61
A day in the life of a rocket scientist 61
What rocket scientists are good at 64
What rocket scientists are not so good at 65
Making a career in rocket science 66

The social dimension 70
Why rocket science? 70
Green space 75
Rocket science serving world peace 77
Rocket science serving world wars 78
The ISS fly-by 80

The moment of truth 81

The crowning experience 82

Glossary 84

INTRODUCTION

Popular comment has it that: 'It doesn't take a rocket scientist to ...', as in 'It doesn't take a rocket scientist to programme a mobile phone.' This is true – it takes a teenager. A rocket scientist would intellectualize the whole process, press two buttons at once, and crash the software. The last person you would want to ask is a rocket scientist.

In fact, if one of your bluffees starts fiddling with his cellphone or two-way pager, neither ignore him nor complain. Instead, draw everybody's attention to him and point out that the first American visitors to the Moon had less computer power at their disposal than a modern Blackberry. They made up for the difference with brains.

“It doesn't take a rocket scientist to programme a mobile phone.' This is true – it takes a teenager.”

If you Google on “It doesn't take a rocket scientist to”, you come up with 150,000 sites. Clearly, rocket science is a major preoccupation in cyberspace. The inference is that the majority of life's mysteries can be solved by ordinary mortals or at least by teenagers, and that only the most elusive problems require the phenomenal brainpower of a rocket scientist. It behoves the bluffer therefore, to be quick to distinguish the rocket scientist from the science itself.

Rocket science is awesome. The German word for 'awesome' is *'verblüffend'*. Note how those clever Germans have discovered the natural connection between rocket science and bluffing. There are people who believe the Apollo 11 Moon landing was one big bluff, an elaborate computer simulation designed by the CIA to fool the world into thinking that America had won the space race against the Soviet Union. For you, the bluffer, it can't get better than that.

In the following chapters, your bluffing skills will be honed to the point where your bluffees swarm and swoon around you, depending on gender. When done, you should be able to paraphrase Betty Midler by claiming at the next ABC (Annual Bluffers' Conference): "I didn't invent bluffing, but I have taken it to new heights."

THE BASICS

Rocket scientists are the people who design satellites and rockets. Some prefer the academic environment where they try to solve unsolvable rocket equations. Others conduct laboratory research to develop new materials which are stronger than steel and lighter than paper. Still more work in industry to cut, bend, mill and extrude the new materials into rocket shapes. A few can be found in the desert or in the jungle preparing the rockets for launch.

Rocket scientists are few and far between, which is one reason why so many problems in the world remain unresolved. More importantly, their dearth offers a great opening for bluffers, since most people don't know what to expect. Next time you are invited out, have your dinner host introduce you to the other guests thus: "Please meet [your good self], the little-known rocket scientist." Then relish their expression of bemused disbelief.

"Rocket scientists are few and far between... and importantly, their dearth offers a great opening for bluffers."

Since rocket scientists are so clever, rocket science must be very difficult. Nothing could be further from the truth. You should modestly insist that a space rocket is just a controlled explosion designed to place a satellite in orbit. The satellite is what matters. Contrary to popular belief, rockets and satellites are different things. The analogy could be made between pregnancy on the one hand, and childbirth on the other. Just as some pregnancies involve twins and triplets, some rockets carry two or even three satellites at a time. Furthermore, it takes approximately nine months between the consummation of a launch agreement and the actual lift-off.

A satellite, on the other hand, requires three years or more to build. Satellites perform lots of clever tasks, ranging from telecommunications, TV broadcasting and navigation all the way to science, remote sensing, meteorology and espionage. Without satellites, there would be:

a. no accurate weather forecasts,
b. no watching satellite TV, and
c. no looking for weapons of mass destruction.

Even though satellites are so much cleverer than rockets, the latter seem to attract more attention. Rockets can be seen and heard, whereas satellites are:

a. hidden away in locked cleanrooms or
b. moving noiselessly through outer space.

Many more millionaires have gone bankrupt trying to develop rockets than satellites. They have overlooked the fact that the operative word in "controlled explosion" is controlled.

As an aspiring bluffer in rocket science, you will easily convince your listeners that putting a burning match to 80 tons of explosives is unlikely to benefit humanity, and least of all the bystanders. Instead, one needs to direct the explosion through something called a nozzle. It looks rather like the inside of your nose, with which you are no doubt intimately familiar. When a person sneezes, the air flow through the narrowest passage of the nose is said to exceed the speed of sound, which explains why an unbridled sneeze will send shockwaves through a crowded space. A rocket nozzle fulfils a similar function by accelerating the combustion gases downward until the rocket begins to move upward (action and reaction). When explaining rocketry to your bluffees, you could equate it to a gigantic sneeze.

"Many more millionaires have gone bankrupt trying to develop rockets than satellites. They have overlooked the fact that the operative word in "controlled explosion" is controlled."

The co-founders of Microsoft, Google, PayPal and Amazon remain unfazed by past rocket flops and are still funding new ventures in space rocketry. You could always tell your acolytes that you are thinking about starting up a rocket company of your own. Thinking aloud doesn't cost money, and it could buy you instant kudos for ingenuity and audacity.

Weightlessness and gravity

Should your talents include launching yourself from a springboard, you will have experienced the exhilaration of weightlessness just before your bum-drop or belly-flop into the water. Of course you are not really weightless because if you were you would be floating into space the moment you took off. But surrendering your body weight to gravity without resistance gives you the illusion of weightlessness.

"The only thing known for certain is that any two physical bodies will attract each other in proportion to their sizes."

Nobody knows what gravity really is, so don't blow your bluffing cover by trying to explain it. The only thing known for certain is that any two physical bodies will attract each other in proportion to their sizes (which fact is best not taken literally by oddly sorted couples).

Some people pay serious money to experience a few moments of weightlessness onboard the Vomit Comet. The V.C. is a modified jetliner that climbs to cruising altitude and takes a nose dive at full throttle to gain speed. Just before hitting the sound barrier, it goes into a steep ascent, throttles back, gently slows down, levels out, almost stalls, falls into another nose dive ... and so forth. Ad nauseam. During this low thrust, so-called 'parabolic flight trajectory', the passengers inside the unfurnished cabin float around as if they were in space, twisting and turning, bouncing against the padded walls and into each other, shrieking with laughter, and trying not to be sick.

“Whichever way one looks at it, overcoming gravity is as important a theme in rocket science as it is in weight-watching.”

The ultimate weightlessness experience is reserved for astronauts and a few well-heeled space tourists who pay $20 million for a week onboard the International Space Station. But, here again, none of them is truly weightless; they are merely being pulled two ways by equal but opposing forces, namely gravity on the one hand, and the centrifugal force on the other hand which comes about from hurtling around the Earth in an orbit.

Whichever way one looks at it, overcoming gravity is as important a theme in rocket science as it is in weight-watching. Since controlling body weight is a popular subject at calorific dinner parties, an alert bluffer will steer the conversation in the direction of the Vomit Comet, describe motion sickness in some detail, and cause the other guests to lose their appetites.

Orbits, trajectories and other encounters

Satellites travel in closed-loop tracks called orbits. Probes follow open-ended trajectories deep into the solar system. People often get satellites and probes mixed up, giving bluffers an entry into the conversation by patiently explaining the difference by a little homily on:

Orbits

Orbits represent a delicate balance between centrifugal forces and gravity. The most popular orbit is the geostationary one. This goes around the equator at a height of 36,000 km (i.e., London to New Zealand and back again) where the satellites revolve around the Earth at the same rate that the planet turns around its own axis. It follows that, to an observer looking up from the Earth's surface, a geostationary satellite appears to stay fixed in the sky day and night (only bluffers are able to actually spot a satellite that far away). This is the perfect orbit for relaying data between points on the Earth without fear that the satellite is suddenly going to disappear behind the horizon.

“There are polar orbits and inclined orbits and sun-synchronous orbits and Molniya orbits [Molniya means ‘lightning’ in Russian]...”

Then there are polar orbits and inclined orbits and sun-synchronous orbits and Molniya orbits

[Molniya means 'lightning' in Russian], all of which are populated by a miscellany of lesser satellites. As a bluffer, you might quote them the same way you drop the names of famous astronauts. This could be productive when trying to impress a date after sunset (or before sunrise), at which time low-flying satellites can be spotted in the sky with the naked eye. Here is a good one: "Honey, see that little star moving across the Belt of Orion? That must be Iridium number 49 in a circular orbit inclined at eighty-six point four degrees on its way to the descending node". It works a treat.

"Trajectories are less repetitive and more adventurous than orbits."

Trajectories

Trajectories are less repetitive and more adventurous than orbits. They are the tracks followed by probes on their way through the solar system. Space may seem weightless at first, but gradually the probes get pulled this way and that as they pass between planets, moons, meteors and other Big Bang spin-offs.

A clever method of exploiting gravitational pull is the so-called swing-by. The launching rocket is rarely strong enough to send a probe all the way into interplanetary space. To make up for the shortfall, the probe latches itself onto the gravitational field of a nearby planet. The speed of the planet is added to the speed of the probe, such that the probe is catapulted into the Universe.

Women love being swept off their feet. When you head for the skating rink next time (as one does), you could give someone's partner a swing-by demo by seizing her around the waist and changing her trajectory with a discus-throwing motion. If you don't have what it takes to scrape her off the wall, experience the feeling by joining her instead on one of those carousels with rotating chairs.

Rocket science in the news

The cost of a space rocket is in the order of 100 million dollars, and the satellite may be worth $200m. About one in 20 rockets fails to place its satellite in the correct orbit. This is known as a launch failure. Some rockets don't even manage to lift off but blow up right there on the spaceport, taking the launch pad with them. Such mishaps make for good TV news coverage. So do the ones where the airborne rocket is destroyed intentionally by telecommand from the ground because it has veered off course and begun to head for the nearest megalopolis. The command prompts an explosive on board the rocket

“Some rockets don't even manage to lift off but blow up right there on the spaceport, taking the launch pad with them.”

to rip open the propellant tanks. The fuel and the oxidizer combust instantly, after which there is not enough left of the rocket to kill a mockingbird.

If the subject of a launch failure comes up, you could say something like: "Looking at the video footage, I'd say that the mixture ratio between the fuel and the oxidizer was off in the left strap-on booster." This pronouncement will instantly generate awe, especially if the ill-fated rocket actually carried strap-on boosters.

A word on strategy: The conversation stopper is the bluffer's emergency brake and should be used sparingly. Nonetheless, if some smart-alec were to challenge your assessment of the accident, remember that asking an impossible counter-question is a tried and tested method of deflection. Try something like: "What did you make of the hypersonic instability in the shock diamonds?"

"Rocket scientists come in two flavours: rocket builders and satellite builders. They are as different as carrots and oranges."

From flaming hell to starlit heaven

Rocket scientists come in two flavours: rocket builders and satellite builders. They are as different as carrots and oranges.

Like their creations, rocket builders make a lot

of noise. Listening to their mumbo-jumbo, one is given the impression that Rockets are Important. This is intentional, yet their only function as you know is to launch satellites into orbit. Even though satellites foot the bill, rocket builders regard their satellite customers as, at best, irrelevant and, at worst, a cursed nuisance. The thing is that satellites get in the way of launching rockets, partly because they don't always fit, and partly because they show up late at the launch site. If the rocket people had it their own way, there would be a launch every day, and it wouldn't really matter if the rocket reached orbit. What matters is the fireworks.

"Even though satellites foot the bill, rocket builders regard their satellite customers as, at best, irrelevant and, at worst, a cursed nuisance."

A space rocket is made up of two, three or four main stages, plus bits of rocketry strapped on to the outside of the lower stage (the strap-on boosters). Each stage is a rocket in its own right. It consists of two big tanks stacked one on top of the other and an engine at the bottom.

One of the tanks contains the fuel, the other the oxidizer. (The reason why a car only carries fuel is that the engine takes its oxidizer from the atmosphere. In space there is no atmosphere, so the rocket has to bring along its own oxidizer. Think of it as a picnic.) Both liquids are so hostile that when they meet and mix in the engine's combustion chamber, they immediately ignite and fight their way supersonically through the engine's nozzle. The laws of action and reaction stipulate that, if the combustion products go one way, the rocket will want to go the other way. You could make analogies with parents and teenagers, or someone turning on the water full blast before grabbing hold of the hose.

On the very top of the rocket sits the satellite. It is mounted inside a heatshield to prevent it from burning to ashes through friction when the rocket ascends through the various layers of the atmosphere. But let us not jump ahead of ourselves. The first thing that happens during a launch is the count-down. Someone in the control centre shows off his ability to count backwards from 10. When he reaches zero, he says "We have a lift-off!" even if the rocket is still on the ground and is about to engulf the entire launch site in flames.

"The laws of action and reaction stipulate that, if the combustion products go one way, the rocket will want to go the other way."

Assuming that the rocket actually lifts off:

1 The first stage and the strap-on boosters provide all the early entertainment as the 100-ton monster gains speed.

2 When that show is over, the second stage ignites and ungraciously sends the first stage back to Earth where it crashes into the ocean or some nomad settlement in Kazakhstan;

3 The third stage treats the burnt-out second stage in a similar manner.

And so forth, until the fireworks display is over.

Somewhere along the way, the atmosphere becomes so thin that the heatshield can be jettisoned (good word, that), thereby exposing the satellite. In fact, the whole launch sequence is one big strip-tease.

Talking through your ears

The satellite in the nude is not a pretty sight. It's no wonder that the last rocket stage spits it out after finishing all the combusting. A satellite, with its eyes and ears mounted skew-whiff on its angular body, looks like a modern junk sculpture. Matters don't improve when it deploys its flimsy appendages, consisting mostly of solar panels and antennas.

Satellites and people have a lot in common. They have eyes and ears, voices and mouths, brains and sensors, and even moods. All of them carry transmitters and receivers so that they can communicate with their masters on the ground. To impress someone at a dinner party with your rocket science credentials, you could try the following analogy.

The big ears sticking out of the sides of a typical satellite are antennas. The bigger the ears, the better the hearing, which explains why people sometimes cup their hands behind their ears. The amazing difference is that satellites also talk through their ears. Here again: the bigger the ears, the louder the voice, which is why we may cup our hands around our mouths (= talking ears) to create a megaphone effect.

"Satellites and people have a lot in common. They have eyes and ears, voices and mouths, brains and sensors, and even moods."

Contrary to popular belief, outer space is far from silent. The biggest enemy of communication is radio noise. If it gets loud enough, the listener can't hear a thing. The most obvious solution is to turn down the noise. You can demonstrate the benefits to your listener by turning to the other dinner guests and yelling: "Shut up, you idiots!!" You and your devotee will now be able to hear a pin drop. Of course, satellites can't just turn around and tell all the other satellites to shut up because that would be rude, but they do try to filter out most of the noise. (NB: when it comes to creating radio noise, the sun is the biggest loudmouth of all, so satellites need to be careful which way they point their antennas. Even microwave ovens and cellphones on the Earth can interfere with satellite communications.)

Another solution to overcome the noise problem is to s-p-e-a-k m-o-r-e c-l-e-a-r-l-y. This may not be so easy after you have had a few drinks, so you should start experimenting as soon as you walk through the door. Your voice is like an unmodulated radio carrier which is modulated by your tongue and lips to form meaningful words. A satellite's radio transmitter works the same way: it produces a constant radio tone (i.e. a carrier) which it modulates electronically with digital bits or analogue waveforms. The radio receiver performs the same function as your brain by demodulating the signal and making sense of the actual message.

“Even microwave ovens and cellphones on the Earth can interfere with satellite communications.”

The cleverness of satellites is of course a mirror image of the ingenuity of satellite builders. Satellite builders are thoughtful, considerate folk who abhor the brashness of their rocket-building counterparts. As a bluffer, you should adopt whichever style suits your personality.

THE GREAT PIONEERS

The greatest names in rocket science are **Sergei Pavlovich Korolev** (pronounced 'karalYoff') and **Wernher von Braun** (pronounced 'fon Brown'). Korolev was Russian, and von Braun was American. Actually, Korolev was born in the Ukraine and von Braun grew up in Germany, but don't rub that in if your audience happens to include Russians or Americans proud of their history. They need every famous name they can get.

"Anybody who is anybody in rocket science has had some dealings with one or the other of the two pioneers."

Anybody who is anybody in rocket science has had some dealings with one or the other of the two pioneers. These anybodies never cease to flaunt their association with fame, and neither should you. But saying that you've been reading up on the two men is not going to create any earthquake at dinner parties – it will be sort of a '1' on the Richter scale. You should at least:

a have met someone who knew them (that's a '3'),
b have worked with that someone ('5'),
c have seen them live from a distance ('7') or
d have worked closely with them for many years ('8').

Remember that each notch on the Richter scale is 10 times the earthquake amplitude of the previous notch.

Should you nonetheless aim for an '8' in the Soviet context, don't refer to him just as Korolev (and certainly not Comrade Korolev) but rather to Sergei ("Sir-gay") Pavlovich. This is how Russians address each other when they wish to show respect, admiration and intimacy. Anybody familiar with the Russian scene knows this, and it saves you from having to explain in great detail how you got to know this giant in rocket science. The fact that he died 40 years ago and you may be only 25 could erode your credibility, so try to look as old and haggard as you can.

"The fact that he died 40 years ago and you may be only 25 could erode your credibility, so try to look as old and haggard as you can."

You would show admiration and intimacy with Wernher von Braun differently. Calling him Wernher Von doesn't quite have the same cachet as Sergei Pavlovich, and "Verne" may get him confused with Jules Verne. Try the German form of respect: Herr Professor Doktor Freiherr von Braun, and point out that "Freiherr" means "Baron".

You are now ready to preamble your anecdotes with: "I once suggested to Sergei Pavlovich that ...", or "The Baron used to tell me about ...". To keep your listeners' attention, it is a good idea to embed your anecdotes in a discussion theme, such as The Power of Competition, or Leaving While You Are Still On Top, or Denying Your Past. Sergei Pavlovich and The Baron are perfect catalysts for all three topics.

Unless you are a surefire improviser, it might help you to know a few facts about the two geniuses.

Wernher Magnus Maximilian Freiherr von Braun (1912–1977)

It is often said that a person never changes; he only becomes more of the same as he gets older. Wernher von Braun is a case in point.

One Sunday afternoon, the 12-year-old von Braun strapped rockets to a cart, lit the fuse, and sent the fire-spitting vehicle careening down a street where the staid burghers of Berlin were taking a leisurely stroll. As the burghers climbed down from the lampposts, von Braun was arrested by the police and spent the next few hours in custody until his politically influential father bailed him out. Instead of punishing their son further, the parents gave him a telescope, hoping that star-gazing would distract the young rascal from further mischief. The therapy only worked while the stars were out, leaving von Braun to pursue his rocket fantasies in daylight for the next 50 years.

> **“The 12-year-old von Braun strapped rockets to a cart, lit the fuse, and sent the fire-spitting vehicle careening down a street.”**

His life-long aim was to send a rocket to the Moon. It doesn't take a rocket scientist to figure out that flying a lunar mission is tricky. Since von Braun was full of tricks, it would only be a matter of time before he succeeded. With a growing track record as a rocket engine designer, he caught the attention of Nazi Germany's political and military leaders. They wanted him to build a stealth missile that could be launched from German soil, fly faster than sound, and strike enemy cities without warning. Money for the project would be in ample supply, especially if he agreed to join the Nazi Party and the SS. Which he did, however reluctantly in hindsight.

“It doesn't take a rocket scientist to figure out that flying a lunar mission is tricky.”

Von Braun saw the assignment as an important step in fulfilling his ultimate dream. Together with his entourage of like-minded patriots, he set up base in Peenemünde on the Baltic coast, from where he launched the first-ever operational long-distance rocket, the V-2. So absorbed was he by his lunar vision that he was thrown in prison for a while, accused of lacking focus on his missile priorities. His preoccupation with the Moon also made him turn a blind eye to the death and destruction that his missiles caused in war-torn Europe. In the 1960s, the American satirical songwriter Tom Lehrer caught the moral irony with the lines:

"Once the rockets are up, who cares where they
come down?
That's not my department," says Wernher von
Braun.

When allied bombers pummeled Peenemünde, the mass production of the V-2 was transferred to a vast underground factory in Nordhausen in Central Germany. Forced labour was obtained from a nearby concentration camp. Throughout World War II, more people died of punishment, exhaustion and starvation while building the V-2 than were killed by its terrifying warheads. All the while, von Braun conveniently kept his gaze steadily fixed on the Moon.

> **"More people died of punishment, exhaustion and starvation while building the V-2 than were killed by its terrifying warheads."**

In the spring of 1945, Russian and American forces were closing in on what was left of Peenemünde. Von Braun and his team debated among themselves to whom they should surrender. Most preferred the Americans and set out on a precarious journey to meet their captors-to-be, all the while dodging the fragmented German security forces who had orders to kill them. It was Wernher's brother Magnus, riding his bicycle, who made contact with the Americans. To his surprise, he found that both they and the Russians were frantically looking for the German rocket scientists. Builders of killer missiles were obviously in great demand, not so much to bring them to justice, as to have them build more of the same.

Von Braun and his team were installed in White Sands in New Mexico, where they helped assemble dozens of captured V-2 missiles for test firing and further improvement. Over the next decade, these improvements culminated in the Redstone intercontinental ballistic missile (ICBM). The rocket performed well during repeated test launches. Just months before the Soviets shook the world with their Sputnik, von Braun applied for permission to add a small rocket stage to the Redstone to launch the first artificial satellite. His request was denied for fear that the Soviets might be offended by an American spacecraft passing over their territory. Instead, during the next test launch, a capsule filled with sand took the place of the extra rocket stage – a political miscalculation that would prove to be both humiliating and enough of a catalyst to make von Braun's dream come true.

“His request was denied. Instead, during the next test launch, a capsule filled with sand took the place of the extra rocket stage.”

Under von Braun's leadership, the United States successfully embarked on a series of space missions involving chimps, humans, moons and planets, but nearly always one step behind the Russians. The

publicity the Americans afforded their upcoming space missions gave the Russians an important propaganda advantage: they were able to upstage the Americans almost every time with even more spectacular missions launched just weeks or days in advance. The Russians, on the other hand, shrouded their space programme in secrecy. If a mission succeeded, the world was taken by surprise – and if it failed, nobody would ever know.*
This is why the Russians seemed unbeatable until, in 1969, Neil Armstrong and Buzz Aldrin finally set foot on the Moon.

“The Russians shrouded their space programme in secrecy. If a mission succeeded, the world was taken by surprise – and if it failed, nobody would ever know.”

Sergei Pavlovich Korolev (1907–1966)

Like von Braun, Korolev spent his youth gazing at the stars, but without the luxury of a telescope. While his schoolmates were force-fed Pushkin's poetry and Tolstoy's epic novels, he was soaking up Tsiolkovsky's theories about rocket propulsion. (In case someone

* Like the time in 1960 when a hundred top military brass gathered around an R-16 space rocket as it was being prepared for its maiden flight. They were all incinerated alive after a technician accidentally sent a telecommand for the second stage to ignite on the pad.

asks, Tsiolkovsky formulated the rocket equation without which rocketry would have been a rocky business.) There was only one way for young Korolev to go, and that was up.

His first move was to build a glider, then another one which was bigger and better, and yet another that pushed the envelope just a tad too far and crashed. Unfazed, he lived to tell the tale and even earned a pilot's licence. Then he set his sights on the Moon.

"During the 1930s, Korolev did succeed in designing and testing a basic rocket engine, but then Stalin suffered a spell of paranoia."

His quest for the Moon was far from plain sailing. During the 1930s, he did succeed in designing and testing a basic rocket engine, but then Stalin suffered a spell of paranoia that also spelled the demise of millions of Russian intellectuals. Korolev was lucky to get away with 6 years of imprisonment, a part of which he spent at the infamous Kolyma labour camp in the Siberian gulag. In their spare time, the prisoners were tortured. Korolev had his jaw broken and his teeth knocked out. Finally he was allowed home to continue designing rocket engines, on condition that he join the Communist Party. In his new job, he was meant to be working on missiles for the Army, but his mind kept drifting back to the Moon.

Korolev quickly realized that liquid propulsion

was the only way to go. And here is your chance to bluff a bit of knowledge about rocketry. Once ignited, a rocket engine that uses solid propellant (space babble for gun powder) will burn continuously until it runs out of the stuff. That's fine for reaching the Moon, but it limits your chances of ever coming back. With liquid propellant, the engine may be switched on and off at will just by opening and closing the flow valves. Now you are able to land on the Moon, turn off the engine, raise your flag, play some golf, and then restart the engine for your return flight.

A minor obstacle facing Korolev was that no-one in Russia had yet fully mastered the liquid propulsion technology, the problem being that the two candidate liquids explode on contact. He learned that, in far-away Germany, von Braun had solved the problem, and the Wehrmacht was busy dispatching his V-2 missiles by air to London, Paris and Antwerp. When Germany lost the war in 1945, Korolev and his fellow rocketeers were sent to the underground V-2 factory to recoup as much hardware, drawings and rocket science that they could fit on the next freight train bound for Moscow. Alas, the Americans had more or less cleaned out the place only hours before the underground site was handed

“Alas, the Americans had more or less cleaned out the place, so Korolev and his team were left to collect the scraps.”

over to the Soviets, so Korolev and his team were left to collect the scraps.

Only a true genius can turn a few scraps of metal into fine rocket science. While von Braun and his merry Peenemünders were playing at fireworks with dozens of complete V-2 missiles in New Mexico, Korolev and his comrades managed not only to build a replica of the V-2, but an improved version at that. Further improvements followed until one day the mighty R-7 intercontinental ballistic missile stood ready for its maiden flight from Baikonur. And it also happened to be capable of placing a satellite in orbit.

“Khrushchev concluded that a satellite crossing the skies of the United States uninvited might earn him a savoury propaganda victory.”

Permission to launch a satellite was initially denied on the grounds that space exploration was a distraction from the more important task of nuking America. But Khrushchev, the Soviet leader best known for banging his shoe on a United Nations conference table, took a broader view. He concluded that a satellite crossing the skies of the United States uninvited might earn him a savoury propaganda victory, so he finally gave the green light. Hence, on 4 October 1957, the very first man-made satellite was carried into orbit by an R-7. The tiny, shiny sphere was called Sputnik, which is Russian for ‘Travelling Companion’.

Khrushchev now with his shoes back on, was dancing with glee over the perplexity that Sputnik was causing around the world. He wanted more of the same, but bigger and better. With one hand he gave Korolev a blank cheque, and with the other he placed a noose around Korolev's neck. Sergei Pavlovich got the picture, for a few weeks later his team successfully launched the first dog Laika (Russian for 'Barker') into space. From then on, Korolev became the driver behind a string of 'firsts' in space: the first man, the first woman, the first lunar impact, the first pictures of the far side of the Moon, the first lunar soft-landing, the first probe sent to Venus, the first orbital rendezvous between two spacecraft, and the first 'extravehicular activity' (space babble for 'spacewalk').

"Korolev's team successfully launched the first dog Laika (Russian for 'Barker') into space. From then on, he became the driver behind a string of 'firsts' in space."

But would he be the first to land humans on the Moon? As a result of his many successes, he was given the go-ahead to design and build a suitable rocket. The result was the gargantuan N-1 featuring five stages, with no less than 30 rocket engines mounted on the first stage. Lifting more than one can carry is a bad career move, and the N-1 spelled the beginning of Korolev's downfall. Over-reacting under pressure, he picked fights with all the wrong people, including Khrushchev and even his own rocket engine designer, who subsequently refused to collaborate on the N-1 project. Korolev's interplanetary probes kept missing their targets, and the stress took a toll on his health. He died while undergoing minor surgery, taking the dream about planting a Soviet hero on the Moon with him to the grave.

"Korolev's name remained a State secret for fear that the Americans might kidnap him."

Soon after Korolev's death in 1966, four of the N-1 monsters were launched in succession, and all four failed miserably. A few more had already been built when the programme was scrapped. Ironically, the left-over rocket engines were sold to the United States, and modern replicas are currently powering some American launch vehicles.

By allowing Korolev to launch Sputnik, Khrushchev had un-leashed the space race with America. But while von Braun was positively glowing in the limelight from his successful rocket launches, Korolev's name remained a State secret for fear that the Americans might kidnap him. It was as if the two pioneers lived on opposite sides of a one-way mirror, where Korolev could see von Braun in all his glory, yet von Braun was unaware of Korolev's very existence.

The space race ended on the day in 1975 when

the Soviet Soyuz capsule docked with the American Apollo spaceship and the cosmonauts shook hands with the astronauts. By then, Korolev had already been dead for a decade.

Paradoxically, while Korolev's status at home skyrocketed after his death, von Braun saw his star decline after the Moon landing. Korolev was given a State funeral, his remains were buried in the Kremlin Wall, statues were raised in his honour, and a major city was named after him.

Edward Mukaka Nkoloso (?–? [unknown])

Few people know how close the Zambians came to beating the Americans to the Moon. It is your duty as a bluffer in rocket science to put the record straight.

"Few people know how close the Zambians came to beating the Americans to the Moon."

Edward Nkoloso [Nnnn-kol-oh-so] was a science teacher who liked to dress in a red and green Superman-like cape. In the 1960s, he soared through the ranks to become the head of Zambia's National Academy of Science, Space Research and Philosophy. One of his philosophical priorities was to send 'afronauts' to Mars, and to plant the Zambian flag on the Moon along the way. Contrary to von Braun and Korolev, he realised that Man was more important than the

Machine, so he quickly recruited 12 aspiring national heroes and subjected them to intensive afronaut training. The 12 included a missionary, a curvaceous teenage girl and her two cats. While employing powerful binoculars to examine the lunar and martian surfaces, they discovered that Mars was populated by primitive natives, and Nkoloso personally extracted an undertaking from the missionary to refrain from converting them to Christianity against their will.

"The training programme consisted of placing the afronauts inside sealed oil drums which were rolled down a hill."

The afronaut training programme took a more pragmatic approach to simulating space travel than the concurrent astronaut and cosmonaut programmes. It consisted of placing the afronauts inside sealed oil drums which were rolled down a hill, thereby giving the inmates a sensation of both swooshing through the vastness of space and hitting the ground on landing. To experience weightlessness first-hand, the afronauts were made to swing on ropes from trees, Tarzan-style, whereby Nkoloso would cut the ropes with a knife at just the right moment to create maximum free-fall.

Having lined up his afronauts, Nkoloso turned his attention to the Machine challenge. He desperately needed funds to develop and build a space rocket. His plea for $19m from the United Nations and for another $1,900m from private investors went unanswered, so he had to settle for the ejection seat of a discarded jet. His plan was to launch the ejection seat from the sports stadium in Lusaka on 24 October 1964, the very day of Zambia's independence from Great Britain. Unfortunately, President Kaunda got involved and told Nkoloso to lower his profile. He reluctantly withdrew from the festivities but kept his Superman cape. And this is how we shall remember him: the indefatigable Superman who combined binocular vision with a profound belief in the equality of men, women, ejection seats and cats.

"He reluctantly withdrew from the festivities but kept his Superman cape."

The moral dimension

The serious bluffer will recognize an opportunity to add a moral dimension to the space race drama. Invite audience participation with a Deep Question: Without the space race between the superpowers, would there ever have been 'a small step for man, a great leap for mankind,' to quote Neil Armstrong? Was it right for von Braun and Korolev to join the Party to advance their ultimate cause? When a candle like Korolev or von Braun is lit, should it be placed under a bushel or in a candle stick? Does anyone want another drink?

WILD ROCKETS

Otrag

Lutz Kayser, a German rocket scientist, was the first to develop a space rocket with exclusively private funding. His vision was to sell rockets for one-tenth of the price of state-funded equivalents. To reach this lofty goal, he formed a company called Otrag and strapped a bunch of glorified chimney pipes together into clusters. The bigger the cluster, the bigger the bang for the Deutschmark, he reasoned with faultless German logic. For example, a customer who wanted a 1-tonne satellite placed in a low orbit would get a rocket made up of 48 chimney pipes in the first stage, 12 in the second stage and 4 in the third stage. By adding several hundred more pipes to the various stages, the rocket would be able to launch a 100-tonne satellite, and more. In America, Wernher von Braun was so impressed by the concept that, after his retirement from NASA, he joined Kayser. With a team like that, only the sky was the limit. Or so they thought.

“The bigger the cluster, the bigger the bang for the Deutschmark, he reasoned with faultless German logic.”

A slight disagreement arose between the two men when it came to choosing a launch site. They agreed that launching the rocket from the heart of Europe might be unwise considering the high population density. Kayser therefore negotiated an agreement with the government of the Congo to set up base in the jungle. Of course, lots of people live in the jungle, but that doesn't seem to have fazed them. What von Braun objected to was the prospect that missile technology might somehow end up in the hands of rogue third-world leaders and be used in warfare.

However, the Congo it was to be. During the 1970s, the rocket testing was so successful that the leaders of the superpowers began to take note. The Soviets and the French did not relish the idea of a German developing long-range missile technology within living memory of the War. The Americans, true to form, did not like the prospect of a German fielding a low-cost alternative to their pricey rocket behemoths. Together, they muscled in on the Congolese government to shut down the launch site.

“They agreed that launching the rocket from the heart of Europe might be unwise considering the high population density.”

No sweat, thought Kayser and moved the whole infrastructure to Libya. Testing continued out in the Sahara desert. In the early 1980s, when a maiden spaceflight looked imminent, the Soviets, the French and the Americans moved in once more, coercing the Libyans to stop Kayser in his tracks, and pressuring the German government to liquidate Otrag for good. The Libyan military happily seized all of Kayser's assets and did what von Braun had feared all along: they tried to continue the rocket development on their own for another decade. Lacking the necessary technical know-how, the Libyans eventually abandoned the project – as did Kayser, having lost $200m in total over 12 years. Here is where you give your thumbs-up (make sure they are your thumbs) for a rocket scientist who thumbed his nose at the superpowers.

“It took some cunning rocket science to make the rotor and the rocket engine spin while the body stayed put, rather than the other way around.”

Rotary rocket

Fast forward to the late 1990s. It doesn't get any wilder than this: a manned space rocket shaped like a giant sugar loaf, with a wildly spinning rocket engine at the bottom, a helicopter rotor on top, and the crew sardined between the fuel and oxidizer tanks. With three separate elements in relative spin motion, it took some cunning rocket science to make the rotor and the rocket engine spin while the body stayed put, rather than the other way around.

Now drop the name **Gary Hudson**, the American inventor of Rotary Rocket. Like Kayser, Hudson had a vision to launch satellites for one-tenth of the going rate. Moreover, he knew how to think outside the box nailed together by the NASA bureaucracy.

The main rocket engines – nearly a hundred of them – were mounted on the periphery of a rotating disc at the bottom of the sugar loaf. The advantage was that the propellant would feed itself from the tanks to the engines by centrifugal force, instead of relying on costly and unreliable turbopumps. The rotation was sustained by slightly canting the engine nozzles sideways.

The helicopter blades on top were folded against the sides of the sugar loaf during the journey into space. When it was time for re-entry into the atmosphere, the blades were deployed and began rotating so as to stabilise the rocket bottom down. When the rocket approached the ground, small rocket engines mounted at the tip of the blades accelerated the rotor, thereby allowing the astronauts to guide the rocket to a soft landing on the desired spot. Then it only took a moment to fill 'er up and take off again.

"When it was time for re-entry into the atmosphere, the blades were deployed and began rotating so as to stabilise the rocket bottom down."

Hudson's rocket was not only cheaper than the conventional, multi-stage rockets, but it could also land anywhere and be re-launched after refuelling, assuming that there was a tanker full of rocket fuel at hand. With its crew of two, it could both inject satellites into orbit and capture stray satellites for repatriation to Earth. The crew complained about not being able to see the ground through the porthole on the side of the rocket, because it made landing rather difficult; but the view into space would have more than made up for the deficiency if the rocket had ever flown that far.

"The crew complained about not being able to see the ground through the porthole, but the view into space would have more than made up for the deficiency."

Which it didn't. After private funding dried up, NASA took over the project, put it back in the box, and sealed it with red tape. But Gary Hudson's name lives on in the annals of rocket science. Who knows ... the idea of a Rocket Plane may pop up again like a Jack in the Box.

Space shuttle

One of the finest examples of rocket science is NASA's Space Shuttle. But is it a rocket, a satellite or an aeroplane? In fact it's all three during various stages of flight, so best call it a "re-useable rocket".

To guarantee maximum safety and reliability, everything inside the Space Shuttle is duplicated or triplicated, except the toilet which is merely complicated.

Space scuttle

Like doctors and lawyers, rocket scientists like to shroud themselves in an air of secrecy. To enhance your image, it's a good idea to 'leak' some classified information – in confidence, of course. (Don't take this advice too literally. The point is to make little-known information in the public domain sound classified.) After a quick glance over your shoulder, you might reveal that, during the heyday of Soviet rocket science, the Russians built a very similar rocket/satellite/aeroplane to the Space Shuttle and called it Buran, which is Russian for 'Thunder'. It was of course bigger and better than the American original, even though (owing to the collapse of the USSR and consequent fall-out) it flew only once, by remote control, without a crew and without a toilet. When the Soviet Union collapsed in 1991, the project was finally scuttled.

“To enhance your image, it's a good idea to 'leak' some classified information – in confidence, of course – (i.e. to make little-known information in the public domain sound classified).”

The toilet question

The space toilet is an excellent subject for a bluffer to raise at dinner parties, not least because it implies intimate knowledge. Invite audience participation by turning the subject into a quiz if you feel so inclined. Here's a good lead-in.

The Shuttle toilet designers were faced with various challenges, the principal ones being:

1. to make it ergonomic for both men and women. After experimenting with flanges and conduits, they solved the problem by making male astronauts sit down, just like their female colleagues.

2. to ensure that, in their weightless state, astronauts of any gender remained seated, come what may. The suggestion that the Shuttle rapidly spin each time someone went to the loo, letting the centrifugal force be a substitute for gravity, was rejected. Rocket scientists were concerned about the safety issue and, had this method been employed when the toilet was located above rather than below the Shuttle's centre of gravity, the centrifugal force would have had the effect of sending the astronaut flying instead of trying.

And the answer? NASA settled for shoulder straps. Lots of them.

CLEVER SATELLITES

Global Positioning System (GPS)

Most people have heard about GPS receivers, the gadgets that remind us where we are and where we are going; but not everyone is aware that the GPS itself is based on a cluster of satellites. For a bluffer in rocket science, GPS is a perfect conversation opportunity.

"24 GPS satellites fly in spaced-out orbits to ensure that four or five of them are visible at any time from any point on the Earth."

Stand up and punch the air with your fists to illustrate how 24 such satellites fly in spaced-out orbits to ensure that four or five of them are visible at any time from any point on the Earth. Only three satellites are needed for a GPS receiver to calculate its own position; the extra ones help improve the position accuracy to a few metres.

Each satellite contains an atomic clock that is accurate down to one gazillionth of a second. The satellite transmits its onboard time to the GPS receiver, which compares the satellite time with its own electronic clock. The time difference between the two clocks, multiplied by the speed of light, yields the precise distance between the GPS satellite and the GPS receiver. With at least four satellites in the field of view, one obtains four distances. Using a four-dimensional form of Pythagoras' theorem, the GPS receiver promptly determines its position in terms of latitude, longitude and height. A doddle for a rocket scientist like yourself, and everybody around you will be nodding with a mixture of understanding, admiration, and fatigue.

As a practical example, describe how a recent climber to the top of Mount Everest used his GPS receiver to establish that the mountain is seven feet taller than previously thought – and that is a long extra climb when you are about to faint from exhaustion and lack of oxygen.

Hubble Space Telescope

Ask your admirers how many times they saw a clear, star-spangled nocturnal sky last year. The answers will probably range from 1 to 5. Ask why and they will answer: “Too many clouds”, or “Too much light pollution”, or “Had more important things to do at night”, or “What does nocturnal mean?”

Claim that if everyone had access to the Hubble Space Telescope they could, as astronomers do, watch the universe unfold night after night. The point is that Hubble orbits the Earth high above the atmosphere, and is therefore unimpeded by clouds, pollution, water vapour, and the dispersal of sunlight that Earthlings call daylight. The lookers are bound to the ground, but the images are relayed via radio signals.

You might also mention that Hubble is an example of rocket scientists slipping up. The original telescope, launched in 1990, had a mirror defect that made it myopic, so the early images were almost unusable. Luckily, the Space Shuttle was on hand three years later to intercept Hubble in orbit and install a monocle to correct the vision. The telescope has remained bright-eyed ever since.

Iridium

Unlike cellular networks on the ground, a satellite serves anybody who lives inside its antenna footprint on the Earth, and a footprint typically extends across thousands of square kilometers. The founders of Iridium succeeded in persuading corporate investors to part with billions of dollars and went ahead deploying 77* satellites to offer global

* Bluffer's note: 77 is the atomic number of the chemical element Iridium; hence the name of the venture, except the number of satellites was slashed to 66 to save money. (66 is the atomic number of Dysprosium, which sounds like a chronic illness, so the name Iridium was retained.)

cellular telephony from space. The funds went to building and launching the satellites, constructing several ground stations, developing a cellphone for consumers, and splurging on publicity. The satellites themselves were a marvel of spacecraft engineering, able as they were to hand over telephone calls to each other and pass them on to neighbouring satellites.

“One had to stand close to a window to make a connection, and calls were interrupted mid-sentence if the active satellite disappeared behind a building or a tree.”

So sure were the Iridium staff of their impending success that their smugness broke all rules of decorum. But no sooner had the satellites been launched than things began to go badly wrong. The cellphones were a year late reaching the shops. Worse yet, the phone was big and heavy like a brick, so an owner needed a briefcase just to carry it around. One had to stand close to a window to make a connection, and calls were interrupted mid-sentence if the active satellite disappeared behind a building or a tree. Voice quality was often scratchy. Worst of all, the phone cost over a thousand dollars, and the price of a call was several dollars per minute. Iridium customers were unimpressed, as were the investors.

At the same time, conventional cellphone networks were expanding much faster than anyone had imagined, and call charges were coming down. A year and $6 billion after the service was launched, Iridium declared bankruptcy, just as its competitors had predicted.

Still, one has to hand it to the Iridium people: they believed in their rocket science and put everything they had on the line. Furthermore the system still exists, having been sold for a pittance by the liquidators. The new owners have solved some of the problems with the handsets, and Iridium now provides satisfactory telephone and data services to (among others) the U.S. military.

"A year and $6 billion after the service was launched, Iridium declared bankruptcy, just as its competitors had predicted."

EXOTIC LAUNCH SITES

Your spellbound bluffees may have heard of Cape Canaveral, Vandenberg, Kourou or Baikonur, so they will of course assume that you have spent time at some of these famous sites. If you haven't, try to divert the discussion to safer territory. With a shrug, give them the impression that if one has seen one rocket launch, one has seen them all, and that the more interesting question is why the launch sites have been placed where they are. Everybody will of course nod vigorously, except for the usual spoil-sport who just nods off.

Tell your listeners that a first-class launch site must fulfil five criteria:

1 be as close as possible to the Equator;
2 not be susceptible to hurricanes and earth-quakes;
3 not oblige rockets to overfly populated areas;
4 be located in a politically stable environment;
5 be free of crocs, alligators and poisonous snakes.

With criteria like that, your audience will infer that there is not a single first-class launch site – a good way of creating tension early in the monologue. But, as always, you have the answers.

The last four items are self-explanatory, but you will need to elaborate on the first one.

Be as close to the equator as possible

An important parameter to distinguish one orbit from another is its inclination, i.e. its slope with respect to the Equator. Thus:

- an equatorial orbit has zero inclination, while
- a polar orbit has an inclination of 90° because it is perpendicular to the Equator.
- the geostationary orbit is equatorial and looks

down on a single geographical area all the time.

A polar orbit covers the entire globe sooner or later, because the Earth rotates underneath the orbit. Orbits with other inclinations have other particular advantages.

One of the miracles of space science is that a satellite's orbital inclination can never be lower than the latitude of the launch site, although it can be higher. Therefore, a satellite launched from Cape Canaveral (latitude 28 degrees) will go into an orbit that is inclined at least 28 degrees. Similarly, a satellite from Baikonur (latitude 48 degrees) will find itself in an orbit inclined at least 48 degrees, not lower. Since equatorial orbits are the most popular ones, all the above launch sites except Kourou (latitude 5 degrees) have a bit of a problem.

"The more the rocket has to divert energy to bend the trajectory, the less energy it has left for lifting the satellite. In space, crookedness comes at a high price."

Luckily, rocket scientists have figured out a way of solving the problem by making the rocket fly along a crooked trajectory. This is a neat nugget to share when you need one. However, the more the rocket has to divert energy to bend the trajectory, the less energy it has left for lifting the satellite. In space, crookedness comes at a high price.

Kourou

Kourou is in French Guiana. It is one of the former three Guianas – the French, the Dutch and the British – at the Caribbean end of the Amazon (except Dutch Guiana is now called Suriname, and the British one is spelled Guyana).

Having given your followers a crash course in geography, tell them that Kourou is the home of the European Ariane launcher – despite not fulfilling criterion No. 5. French Guiana is known for its delightful Caribbean cuisine that includes anaconda snake and cayman alligator. The native Indians live in idyllic hamlets where a dwelling may consist of four wooden poles in the ground with a roof of palm fronds. The men hunt for food while their womenfolk spin cotton, pick avocados and make pottery using local raw materials. Despite their bucolic lifestyle, the villagers get a raw deal due to being rattled by thunderous rocket launches.

“Kourou is in French Guiana which is known for its delightful Caribbean cuisine that includes anaconda snake and cayman alligator.”

However, because reliability and accuracy count (and they do to satellite owners), Ariane is one of the most successful commercial space rockets. It is also unique in that it routinely launches two satellites at a time to bring down the cost. Soon Kourou will be the home of a smaller Italian rocket called

Vega, and the Russian workhorse Soyuz will move from Baikonur to Kourou to take advantage of the near-equatorial location. If the Indians have been rattled by Ariane, they ain't seen nothin' yet.

Cape Canaveral and Vandenberg

Cape Canaveral in Florida, from where the Apollo missions to the Moon lifted off in the 1960s and 70s, is an alligator-infested swamp. Nowadays it is the home of the legendary rockets Atlas and Delta, as well as the Space Shuttle. Most of them launch eastward over the Atlantic Ocean, thereby getting a free additional push by the Earth's own rotation.

Atlas and Delta are also launched from **Vandenberg**, situated along a pristine Pacific coastline in California. Launching north or east is ruled out because of conurbations. Westward launches are awkward since they go against the rotational movement of the Earth; so that leaves launching south, and this is why most satellites from Vandenberg go into polar orbits.

Baikonur

The immense Baikonur cosmodrome, scattered across the bleak rubble of Kazakhstan, used to be the prime launch site of the Soviet Union. It is actually situated near the railway hamlet of **Tyura-Tam**, while the real Baikonur lies 220 km to the north. The Soviets used ≠'Baikonur' as a decoy lest the Americans should decide to erase the cosmodrome from the map.

"The Soviets used 'Baikonur' as a decoy lest the Americans should decide to erase the cosmodrome from the map."

The best-known rockets going up from Baikonur are Proton for commercial and military launches, and Soyuz to ferry astronauts and supplies to the International Space Station.

After the collapse of the Soviet Union, the Kazakhs took over the launch complex, and Russia now pays an annual lease to continue using it. The relationship between Russia and Kazakhstan remains rocky, and it is not unusual for the local authority to turn off the electricity during terminal countdown.

Truth or Consequences

A gem for your bluffing endeavours in rocket science, a new launch site is being built in New Mexico to accommodate Virgin Galactic's space tourism venture in the near future. The site is located near the town of **Truth or Conse-**

quences. For the superstitious, linking an audacious space venture with a town thus named may seem ominous, the more so when considering how the name came about.

> **“For the superstitious, linking an audacious space venture with a town named Truth or Consequences may seem ominous.”**

Truth or Consequences was a popular American radio and TV quiz show that ran for almost 40 years. In 1950, host Ralph Edwards promised to anchor it to the first town that agreed to rename itself after the name of the show. A spa resort called Hot Springs in New Mexico rose to the challenge, and every year thereafter, Ralph Edwards kicked off his annual show season there with parades and beauty contests. The idea of the show was to mix the original quiz element of game shows with wacky stunts.

Say no more.

But there is more

The only launch site that meets all five criteria is a converted oil platform called **Ocean Odyssey**. It propels itself from its home base in San Diego to a point on the Equator in the middle of the Pacific Ocean. The platform and the attendant command ship carry the rocket, the control centre and all the other paraphernalia that goes with a launch site.

Not surprisingly, the rocket is called **Sea Launch**.

A brilliant solution you might think, because not only does the site allow orbits to be inclined at will, but also there are no earthquakes, no hurricanes, no people living downrange, no political situation whatsoever, and no snakes, crocs, or alligators. The only snag with the oil platform is that it takes two weeks for it to make the journey to the Equator, and another two weeks to get back to home base for refurbishment and reloading, so the launch rate is less than optimum.

> **“Not only does the site allow orbits to be inclined at will, but also there are no earthquakes, no hurricanes, no people living downrange and no snakes, crocs, or alligators.”**

Sea Launch is a model of international co-operation. The platform was built by the Norwegians, the launch rocket is a modified Ukrainian-Russian Zenith and Boeing of the United States is in charge of marketing and sales. And the rocket actually works, most of the time.

Just in case someone points out that there are also launch sites in China, you can nip that intervention in the bud by adding “At least three in China, plus several others in Russia, India, Japan, Israel and Brazil. Any more questions?”

MOVERS AND SHAKERS IN SPACE

The customer and the contractor

Anything that flies into space is a spacecraft, so that includes rockets and satellites. For every new spacecraft, there is a Customer and a Contractor. The Customer has lots of money and opinions. The Contractor wants the money and tries to figure out how to separate it from the opinions. Hence a marriage of convenience is concluded. Like most marriages, it is a roller-coaster ride.

"Just as a suitor goes looking for a suitee, the Customer starts his search for a Contractor. From then on, it's all about bluffing."

Designing and building a new spacecraft requires upwards of three years, i.e., about the same time it takes many people to get engaged, married, honeymooned and baby-hatched. First of all, someone has to decide that it's time to get married, which makes him or her a suitor. In the context of rocket science, the suitor is called the Customer. Just as a suitor goes looking for a suitee, the Customer starts his search for a Contractor. From then on, it's all about bluffing plus a little bit of rational negotiation, so it is right up your street.

For the sake of linguistic simplicity, assume that the suitor is a man and that the object of his dreams is a woman. This assumption fits well in the space context, where Customers come across as aggressive and opportunistic, while Contractors are unpredictable and like to play hard-to-get. Usually, the suitor has some idea what his perfect partner should look like, and also what she should be capable of, so the next step is to have a look around. In parallel, the Customer puts himself on display at conferences or sends out love letters in the form of calls for tender to potential bidders. If he is influential and has lots of money, the bidders will come running. The most accommodating bidder is likely to receive a ring from the Customer.

The ensuing negotiations may be likened to a blissfully blufferous engagement period. While the Customer pretends to be generous, fair and reasonable, the Contractor promises to be faithful and punctual. The wedding is witnessed by a motley crowd of lawyers, accountants, rocket scientists and marketing types who do a lot of back-slapping over bottles of champagne. The Customer and the Contractor enjoy a short honeymoon, during which the Contractor had better become pregnant. But will it be a rocket or a satellite? Here is where the analogy with marriage comes apart, because the choice is already stipulated in the contract. There may be many surprises when a

“While the Customer pretends to be generous, fair and reasonable, the Contractor promises to be faithful and punctual.”

spacecraft is born, but gender isn't one of them.

Three years later, the newborn is placed in a cleanroom and is carefully checked out by rocket scientists in white coats who make sure that the electronic equivalents of arms, legs, eyes, ears, voice, liver, kidneys, lungs, brain and heart all work immaculately. If they don't, a quick transplant solves the problem.

> **"Three years later, the newborn is placed in a cleanroom and is carefully checked out by rocket scientists in white coats."**

When all systems are go, the spacecraft is launched. Depending on the outcome, the Customer and the Contractor may decide to have another baby, albeit not necessarily with each other.

Who is who in the space business

It is a boon for the bluffer that bluffing plays such a central role in the partnership between Customers and Contractors, and ultimately in the creation of rockets and satellites. Exploit it to the full. Strengthen your credentials by painting an intimate picture of the Customers and Contractors, as if you had spent most of your career wandering their carpeted corporate corridors.

A good beginning is to drop the names of the biggest Customers. You have probably already mentioned NASA so many times that the edges are getting a little blunt. For a change, try ESA (pronounced Eees-ah) which stands for the European Space Agency, or l'Agence spatiale européenne if you want to show off. ESA is an intergovernmental research organisation employing hundreds of rocket scientists who thrive in the cosmopolitan working environment and enjoy tax-free salaries.

Tell your bluffees that the ESA people are very good at rocket science. Not only do their rockets send probes to other planets, but they have even managed to park a few at the 'Lagrange' points.

Before dropping more names, you should mention that there are two kinds of Customers, namely those devoted to science (i.e. spending money) and those

Lagrange points

Don't let the blank stares from all directions discourage you. Explain that these are points in space where the gravitational forces of the Earth and the Moon balance out with the centrifugal force, such that a probe inserted at that point will stay put. Staying put helps probes to contemplate outer space without spending precious rocket fuel.

There are five Lagrange points in total, but don't push your luck by trying to explain why.

serving humanity (i.e. making money and making war). NASA and ESA belong to the former category. Between them, they spend billions of taxpayer dollars every year. Among the latter, you find the big telecommunication operators like Intelsat, Eutelsat, Inmarsat and SES Astra. Rather than spending your tax money, they make you part with your disposable income to provide you with direct-to-home TV, international telephony, and various data relay services.

> **“The military commission spacecraft that spy and eavesdrop on humanity with a fervor that makes Orwell's *Nineteen Eighty-Four* seem like the height of privacy.”**

The biggest Customers these days are the military. They commission spacecraft that spy and eavesdrop on humanity with a fervor that makes Orwell's *Nineteen Eighty-Four* seem like the height of privacy.

On the Contractor side of the business one finds aerospace giants like Boeing, Loral and Lockheed Martin in the United States, and EADS and Thales Alenia in Europe. In Russia, state-owned Khrunichev has soaked up most of the former Soviet spacecraft manufacturers. All these Contractors build satellites, and half of them make rockets as well.

The relationship between Customers and Contractors is based on bluff. The biggest bluff of all is the suggestion that space contracts constitute a 'win-win deal' for both parties. It doesn't take a rocket scientist to figure out that for every winner there must be a loser. As a supreme bluffer, don't let yourself be persuaded otherwise.

The Coca-colonisation of space

Space tourism is a very different kind of business from building spacecraft, but it is a space business all the same. Because it receives ever more publicity, the subject is bound to come up at one of your bluffing sessions. Your bluffees may have read about Dennis Tito, Mark Shuttleworth, Gregory Olsen, Charles Simonyi or (the first woman) Anousheh Ansari who each paid $20 million for a 10-day stay at the **International Space Station***. People with that kind of cash are known to spend it on things like a 300-foot yacht equipped with a helicopter pad and three satellite antennas. Crazy things, but at least they have a tangible value. The residual value of a trip into space is more difficult to measure.

Not so, say those who have done it – and you should sound as if you are thinking of doing the

* A sprawling, orbiting space complex that looks like a giant man-eating dragonfly. It should always be referred to as "the ISS".

same. Apparently the view of the planet is so moving and unforgettable that even the most hardened astronaut melts like ice cream at the sight. Given an opportunity, these same space tourists would happily put down another $20m for one more glimpse.

Ironically, NASA stonewalled (American) Dennis Tito, when he first approached the Administration with his idea to pay privately for a ride to the ISS. "The Station is for professionals," they declared, "not for amateur space enthusiasts." Unfazed, Tito then turned to the Russians who are senior partners in the ISS mission. They were, as always, short of cash for meeting their partnership obligations, and therefore welcomed Tito and his chequebook with open arms. Having missed the boat, NASA tried to block his access with reference to the ISS treaty which only talks about professional astronauts and cosmonauts. So the Russians had Tito undergo full cosmonaut training at their Star City centre and, before long, he was shaking gloves with the American astronauts onboard the ISS.

"NASA stonewalled (American) Dennis Tito, when he first approached the Administration with his idea to pay privately for a ride to the ISS."

Fuelled by the boundless enthusiasm of well-heeled space tourists, several entrepreneurial companies have risen out of nowhere to offer more... and less. The 'more' category includes a trip around the Moon (at $100m per seat, meals included) and eventually to Mars (price to be announced). All that is needed is a suitable space capsule, enough physical and psychological training to survive weeks, months and years inside a small tin can, and an even fatter chequebook.

"The 'more' category includes a trip around the Moon (at $100m per seat, meals included) and eventually to Mars (price to be announced)."

The 'less' people are taking a more down-to-earth approach by offering a few minutes of weightlessness in suborbital flight – meaning that the spacecraft reaches space but does not go into orbit before falling back to Earth. Some of the latter are quite serious and credible, like Richard Branson's Virgin Galactic, set to begin commercial passenger launches from Truth or Consequences in 2010.

The human factor

Apropos space tourism, there is a camp in the rocket science community that believes human beings have no business flying in space. To make their case, they cite motion sickness, osteoporosis, and muscular atrophy. One is reminded of the arguments once raised against Icarus, Columbus and the Wright Brothers when they were perceived to defy the laws of Nature. To make matters worse for rocket scientists, modern astronauts are not as expendable as the early pioneers, so the craft in which they travel has to be equipped with luxuries like air, water, parachutes and Coca-Cola. This is why many in the rocket science community prefer to launch robots which can be guided by joystick from Mission Control.

"Modern astronauts are not as expendable as the early pioneers, so the craft in which they travel has to be equipped with luxuries like air, parachutes and Coca-Cola."

To bluff your way into a discussion on the above topic, you might add that humans are slow thinkers compared to computers, and what trickle of thoughts they do produce is corrupted by Moods, Feelings and Opinions. But there is one thing humans have which is lacking in computers, namely Judgment. The theory is that there are situations in space where on-the-spot Judgment is useful, for instance when plugging a leak or putting out a fire onboard a space station, when replacing a telescope lens in the Hubble Space Telescope, or when freeing a stuck solar panel on a power-starved communications satellite. This is the kind of moving and shaking where humans outperform robots.

Someone might object that if humans stayed away from space, there would be no use for space stations in the first place, and therefore no need to plug holes and put out fires. You could reply that this merely proves your point – and then make yourself scarce while your bluffees ponder your logic.

Feel free to show off your inside knowledge of rocket science and foreign languages by reminding everybody that American space flyers are called astronauts, while the Russians call theirs cosmonauts, and the Chinese variety is known as taikonauts (from *taikong*, meaning space). A person volunteering to go to the Moon is a lunatic.

“Rumour has it that, in the early days of space travel, American male astronauts had to be married. There was a reason for this.”

Rumour has it that, in the early days of space travel, American male astronauts had to be married, whereas their female colleagues could be married, single, or anything in between. There was a reason for this. Most male astronauts have made their careers in the U.S. Air Force as fighter pilots. As such, they were taught to use their brains and make split-second decisions on their own initiative. None of that is of any use during human space flight, since all important decisions are taken by Mission Control in Houston and the astronauts are expected to implement them without argument. The purported reason why NASA required male astronauts to be married is that only married men are used to obeying orders.

TO BE OR NOT TO BE A ROCKET SCIENTIST

This is a subject that preoccupies many young people. By offering career advice, you demonstrate that you are an Elder Statesman in the profession. A good introduction is to describe a typical day in the life of a rocket scientist.

A day in the life of a rocket scientist

07.58 Rocket Scientist (typically a 40-year-old male) arrives at the Research Institute in his 4.2-litre, red, low-slung, midlife-crisis vehicle with the roof down. He is early as usual, not so as to get an early start on rocket research, but to beat his fellow scientists to the few remaining slots in the parking lot. He disentangles his knees from under his nose and leaves the vehicle.

08.12 Approaches the main entrance. Elbows his way past colleagues who are having an early smoke.

08.17 Sits down at his desk and switches on the computer. The screen says: 'Weekly computer maintenance in progress to serve you better. Your friendly IT Department.' To allay boredom he begins to defragment the hard drive.

08.55 Goes to the conference room to chair the weekly co-ordination meeting with the team. He has set it to begin at 08.55 rather than 09.00, because in his latest management course he was told that it grabs people's attention and makes them arrive on time.

09.15 The first team members arrive for the meeting.

10.23 Closes the meeting and goes upstairs to join the department heads who are discussing how to enhance the Institute's image among the media, the general public and especially the politicians who approve the annual budget. It is decided to switch from steel paperclips to brass ones with immediate effect.

12.19 Goes to the cafeteria for a light lunch consisting of rocket salad.

13.48 Wanders from the cafeteria to the satellite checkout area. Dons a white gown and a white hat and white gloves and slippers. Looking like a snowman, he enters the cleanroom where the Pluto Lander is being prepared for shipment to the launch site. The probe is so dazzlingly beautiful with its silver mirrors and golden blankets that he has eyes for nothing else. This is why he stumbles on the umbilical cable on the floor and accidentally pulls it out of its connector on the satellite. The technicians walk off the job, because union rules state that only union members are allowed to pull the plug.

“The probe is so dazzlingly beautiful with its silver mirrors and golden blankets that he has eyes for nothing else.”

14.01 Spends the next hour apologizing to management and union leaders for his clumsiness down in the cleanroom. The boss says he is guilty of having caused another two-hour delay in the 3-year development programme, at great cost to the taxpayer.

15.01 Stressed out, he goes back to his office and asks himself what he is doing here. He is cheered up by a framed quote from Wernher von Braun on his desk: “Basic research is what I am doing when I don’t know what I am doing.”

16.55 Leaves the Institute to go home after so much satisfying research. Can’t find his low-slung car in the vast parking lot because he forgot where he put it, so he has to wait until everybody else has gone home. When he does find it, he opens the door to let out 20 gallons of rain water, sits down behind the wheel, aligns his knees under his nose, slides the lever of the automatic to D, and drives off into the sunset.

17.48 Rediscovers sanity in the embrace of his family. Wife asks how his day went. He lies. He asks their two teenagers how their exams went. They lie. Watches government ministers on TV trying to justify the war in Iraq. They lie. How relaxing, he thinks, not to be always on the look-out for the Truth.

00.44 Receives the Nobel Prize for his contributions to Rocket Science from the King of Sweden. Dreams don't get much sweeter than this.

What rocket scientists are good at

- Chemistry – to establish what mixture of rocket propellants will give the biggest bang.
- Mathematics – to predict what will happen after the fuse has been lit.
- Propulsion – to run as fast as possible from the launch pad after the fuse has been lit.
- Physics – to explain in hindsight what really happened after the fuse was lit.

- Mechanics – to make sure the rocket doesn't collapse as it accelerates after lift-off.
- Electronics – to build computers that think fast enough to second-guess what the rocket is up to.
- Control systems – to keep the rocket on the straight and narrow.

What rocket scientists are not so good at

- Reading books for pleasure, unless they contain equations.
- Putting two sentences together.
- World politics (though some notables will join whichever political party promotes their cause).
- Predicting the ultimate cost of new space ventures.
- Understanding the meaning of deadlines.
- Programming mobile telephones.

Making a career in rocket science

Have the students repeat after you: Being a rocket scientist is challenging, rewarding and plain fun. The ultimate aim is to seek the Truth by questioning what everybody else has ever done. But it is also stressful because of the huge risks and values involved. Burn-out occurs around 45 – and we are not talking rocket stages here.

Students of rocket science learn from those who have just graduated that career-making begins with a seemingly endless vicious circle: one doesn't get hired unless one has experience, but one won't get any experience unless one is hired. Your advice as an expert bluffer is for the student to get his or her foot in through the back door.

"Students of rocket science learn from those who have just graduated that career-making begins with a seemingly endless vicious circle."

University students often wedge their way in by offering to do their summer apprenticeship or write their graduation thesis on a research subject chosen by a favoured space company or organisation. The student is paid just enough to cover local housing and subsistence expenses, bearing in mind that the hosting entity is probably located in a foreign country. Although the neophyte will get away with English, speaking the local language helps getting the foot even deeper into the crack between the door and the jamb.

Having gained some experience, the student may be invited to join the permanent staff after graduation. There could hardly be a greater incentive to get those last exams out of the way. From then on, the career path can be rapid, depending on the employer and, surprisingly, on oneself.

You should mention that many young space employees soon find themselves at a juncture in their careers where a choice must be made: become an astronaut, go deeper into science and engineering, or accept a management post.

To become a male astronaut, one has to have a Ph.D., be a fighter pilot, possess all one's faculties, be immune to physical and mental torture, and be married. Female astronauts need not be married but must still endure torture.

A more probable career path is found in science or engineering, where much professional satisfaction comes from being a renowned authority in a particular field, and being invited to give papers at international conferences attended by thousands of like-minded nerds.

> **“Paradoxically, those who excel the most in space science or engineering are often promoted into management.”**

Paradoxically, those who excel the most in space science or engineering are often promoted into management. Not only are their former colleagues deprived of their knowledge and wisdom, but the promotees often lack any discernible management talent. The standard remedy is to dispatch them to management courses where they learn all about hygiene factors and motivation dynamics. Thus armed, they embark on a dog-eat-dog existence where they either eat or get eaten. The pay is good, but burn-out is never far away, and nothing tastes worse than burnt dog-meat.

Those who eat the most dogs thrive and become Directors and Chief Executive Officers. When travelling, the CEO flies in First Class and the Directors in Business Class, while the rest of the staff are seated in the nappy-changing area in the tail section. Some space employers stipulate that the CEO must fly alone in case the plane crashes which would be very bad for rocket science. Up to three directors may travel on a given flight, since each is worth one-third of a CEO. There may be any number of ordinary staff on a flight, because the tail section is the most likely to remain intact in a crash.

"The pay is good, but burn-out is never far away, and nothing tastes worse than burnt dog-meat."

A few select canine carnivores become government ministers in charge of astronomical space budgets and the destiny of thousands of space workers. Titles are very important at that level. In America, the Land of the Classless, one might find oneself elevated to the rank of Acting Deputy Assistant Undersecretary for, say, Rocket Science.

(In this context, Secretary is not someone who makes coffee for the boss, but is a person at ministerial rank.)

This is good. An Undersecretary is, by definition, an underling. The Assistant carries the Undersecretary's briefcase on business trips. This is OK because it's not heavy. The Deputy is a person who is sometimes allowed to stand in for the Assistant. This is less good. Being the Acting Deputy is a temporary honour after the Deputy either died recently or was fired. The Acting person is not trusted with the responsibility of ever taking over the job; he is just a stand-in, a walk-on, an extra. This stinks. And yet the Acting Deputy Assistant Undersecretary for Space Science is way above the CEO in rank, and the CEO is head and shoulders above the young space worker about to make his or her most important career choice.

> **"The Assistant carries the Undersecretary's briefcase on business trips. This is OK because it's not heavy."**

Your sagacious advice to the younger generation: don't be blinded by titles. Choose a career ladder that points upward, rather than one that curves itself to a treadmill or tips up like a see-saw. Or else forget about the career ladder altogether and just enjoy rocket science for all it is worth.

THE SOCIAL DIMENSION

More money is invested in rocket science every year than in almost any other human endeavour. Now that you have bluffed yourself into the role of spokesperson for the profession, you are likely to be challenged by disgruntled taxpayers – and there are many of these. If you happen to be pushed into one of the following corners, here are some emergency exits.

Why rocket science?

There will always be a spoilsport among your bluffees who thinks space exploration is a waste of funds. For that kind of money, one could build umpteen new hospitals every year, he or she will insist, not to mention filling the stomachs of a billion poor people around the world. The inference is that, without the costly distraction of space exploration, the world would be a happier place.

“More money is invested in rocket science every year than in almost any other human endeavour.”

The argument has merit, so don't lose credibility by dismissing it out of hand. It is true that, for the price of one large satellite, one could build quite a respectable hospital. The attendant launch cost would cover the purchase of a thousand new ambulances. There might even be money left over for salaries.

Now to the counter-arguments. Whoever questions the value of space exploration is also questioning the virtue of Columbus discovering America. But this is a contentious issue these days, so best avoided. You could try "The pursuit of science is what distinguishes humans from the apes", but at the rate humans are killing off the apes, there soon won't be any left to distinguish themselves from, so maybe this rebuttal should be kept on the backburner as well. A more promising approach for the bluffer is to move from philosophy to hard facts.

"Whoever questions the value of space exploration is also questioning the virtue of Columbus discovering America."

Hard fact 1

Only about half the space exploration budget comes from the taxpayer, namely the funds earmarked for space science and military applications. Almost everybody agrees that the military are doing a sterling job spying and eavesdropping on people who hold opposing views. So there go half the ambulances, and half of the new hospital wards will have to be closed.

Hard fact 2

The other half of the investment in space exploration (space exploitation is a better word) is undertaken by commercial applications, such as TV distribution, CD-quality radio, Internet access in remote areas, and taking pictures of the planet. These are all financially self-supporting through subscriptions and the sale of user terminals (give or take a few spectacular bankruptcies).

Hard fact 3

There are more useful spin-offs from space technology than your doubters can count on their collective toes and fingers. A web site called 'Ultimate Space Place' enumerates all manner of them, and here are some of the important ones:

Water Purification – Using electrodes that generate silver and copper ions to kill bacteria and algae in municipal swimming-pools without the need for chemicals. Keeps bathing suits from losing their colour, and grey hair from turning green.

Tollbooth Purification – A laminar airflow technique used at tollbooths on bridges and turnpikes to decrease the toll collectors' inhalation of exhaust fumes and spare them the agony of slow suffocation. (Don't even think the thought ...)

Laser Angioplasty – Using a 'cool' type of laser which preserves blood vessel walls and offers precise non-surgical cleanings of clogged arteries with extraordinary precision and fewer complications than in balloon angioplasty. No more need to feel guilty about eating those greasy fried eggs and bangers for breakfast.

Emergency Response Robot – A derivative of various Moon and Mars rovers that checks for bombs in abandoned vehicles and suitcases.

Scratch-Resistant Lenses – Coating the plastic lenses of spectacles with a film of diamond-like carbon that not only provides scratch resistance, but also decreases surface friction, thereby reducing water spots. A big improvement over the Japanese invention of spectacles equipped with electric windshield wipers.

Golf Ball Aerodynamics – Employing NASA aerodynamics technology to create a pitted ball surface that sustains initial velocity longer and produces a more stable ball flight for better accuracy and distance. (Before some clever clogs gets in first, say it's not obvious how this can be a spin-off, since there is no atmosphere and hence no aerodynamics in space.)

Personal Storm Warning System – A lightning detector that gives a 30-minute storm warning to golfers, boaters, homeowners, private pilots, and philandering spouses.

Programmable pacemaker – Incorporating multiple NASA technologies, the system consists of the heart implant and a physician's computer console containing the programming and a data printer. The physician communicates with the patient's heart through wireless telemetry signals, like a space-age Cupid.

Athletic shoes – Moon Boot material encapsulated in running shoe midsoles improves shock absorption. In the near-weightlessness of the Moon, shock absorption was of course a major preoccupation for the astronauts.

Self-righting life raft – Developed for the Apollo program, it fully inflates in 12 seconds and protects lives in dire weather conditions. May prove useful for paddling across the lunar Sea of Tranquillity in bad weather.

Virtual Reality – The kind where one walks around wearing a helmet with an internal TV screen. The screen projects a pre-programmed 'reality' whose scenes change with the turning of one's head, just like in real life. One can even see one's hands and feet against the virtual background. Friends can be made to appear in the form of avatars who agree with everything one says and does. Great for stand-up sex.

Green space

Sooner or later, you will be confronted by some tree-hugger who wants to know what rocket scientists are doing to keep space green. The question is valid, so nod pensively as if your interlocutor had just displayed an intellect almost equal to your own.

Given the relative velocities among objects in space, even a small flake of paint from some abandoned rocket is enough to create a deep crater in the windshield of the Space Shuttle. A bigger piece could puncture the fuselage.

Luckily, the Earth's atmosphere is a good vacuum cleaner of space debris. Though the air is too thin for breathing comfort even at 8 km (5 miles), it extends several hundred kilometers out into space and has the effect of slowing down low-flying satellites. As they slow down, they lose height, which makes them dive deeper into the air and slow down even further... and so forth until they re-enter the atmosphere and burn up.

"Space is self-cleaning up to an altitude of about 600 km (the distance from London to Paris). Beyond that, satellites will hang around for all eternity."

So space is self-cleaning up to an altitude of about 600 km (the distance from London to Paris). Beyond that, satellites will hang around for all eternity unless something is done about them. That something consists in firing a retro-rocket onboard the satellite (if it has one), thereby slowing it down artificially and making it re-enter the atmosphere. All 66 Iridium satellites and their spares, flying at an altitude of 840 km, were slated to be taken out of orbit in this manner after the original company went bankrupt. It was only the eleventh-hour intervention by an investor that prevented this mind-boggling act from being committed.

However, firing retro-rockets won't do the trick for satellites way out in the geostationary orbit. So popular is this orbit that hundreds of satellites perch on it like squawking parrots, and there is a serious shortage of elbow room.

"So popular is this orbit that hundreds of satellites perch on it like squawking parrots, and there is a serious shortage of elbow room."

Because these nominally stationary satellites become restive under the influence of gravity gradients (bluffers, sharpen your pencils) and especially luni-solar perturbations (bluffers, take note), there is a risk of collision. Tell anyone who is still listening that a collision would in fact be inevitable, if not imminent, unless something was done about it. Were a collision to occur, the violence of the encounter would make both satellites disintegrate into bullet-sized fragments, each of which would be capable of splitting other satellites further down the track. A chain reaction might ensue and, before long, the Earth would find itself surrounded by a Saturn-like ring consisting exclusively of man-made debris.

Here is where you tap your finger against your temple to indicate that rocket scientists think of everything. Just before a geostationary satellite runs out of propellant, it fires its rockets one last time to propel itself out of the precious geostationary orbit and into a so-called graveyard orbit. And there it will remain forever and ever, deaf and mute, colliding as much as it wants with other space relics, and yielding bits of 21st-century technological finds for the enjoyment of visiting extraterrestrials.

“A chain reaction might ensue and, before long, the Earth would find itself surrounded by a Saturn-like ring consisting exclusively of man-made debris.”

Rocket science serving World Peace

A true bluffer in rocket science is a purported expert not only in physics and linguistics, but also in international conflict and co-operation. A compelling example is the paradox that the race to space has stimulated both fierce competition and fine collaboration among nations.

Point out that the successful 1969 moon landing, albeit a small step for Man, was indeed a great leap for Mankind. (Rocket scientists love such gushy, pseudo-philosophical phrases. Space babble is riddled with bombastic expressions like 'Planet Earth', 'Our Ultimate Origins' and 'The Farthest Reaches of the Universe'. Insist that the landing might never have happened were it not for the competition between the superpowers of the day.

Gradually, the competition matured into co-operation as American and Soviet spaceships were made to dock with each other. Later on, cosmonauts were invited to fly onboard American spacecraft, and vice versa.

> **"Space rockets often have their origins in deadly missiles. The European Ariane and the Ukrainian Zenith are notable exceptions."**

Mention that, in the new co-operative spirit, the Americans refer to space as a 'level playing field' which they try to tilt ever so slightly in their own favour, because they want to be the first to land on Mars.

Rocket science serving World Wars

As you have already hinted when introducing your pals von Braun and Korolev, space rockets often have their origins in deadly missiles. The German V-2 evolved into the Soviet R-7, Semyorka (which means 'Little 7'), and the American Redstone, both of which were small potato compared to the ICBMs that gave birth to the likes of Atlas, Delta, Soyuz and Proton. The European Ariane and the Ukrainian Zenith are notable exceptions, having been designed from the outset for the Peaceful Uses of Outer Space, in the spirit of the U.N. treaty bearing that name.

Be that as it may: no bluffing about rocket science would be complete without mentioning missiles. The cruise missile is a favourite with the media. It has enough eyes and brains to know where it is, compare its position with the topographical maps in its memory, and head straight for the preprogrammed target. Without missiles, attackers and defenders would still be on equal turf, and no progress whatsoever would have been made in the killing business since the Second World War.

“The cruise missile is a favourite with the media. It has enough eyes and brains to know where it is.”

There are basically two kinds of missiles: **guided** ones and **ballistic** ones.

1. Guided missiles have wings and keep their engines running. Like aeroplanes, they rely on the atmosphere for steering.

2. Long-range ballistic missiles have no wings and spend most of their journey in unpowered flight. They reach outer space before dropping down on their targets, so in that sense they are space-

The ISS fly-by

If your toilet quiz hasn't seen you evicted from the party, your next gambit could be to invite the other guests to view the ISS with their own eyes. Just check the NASA web site on the Internet before the party to find out the time of the evening's pass above your neck of the woods. Also make sure to memorise where on the horizon the ISS will first appear.

Five minutes before the stated time, announce to the guests that you have arranged with NASA to have a special fly-by in honour of your generous hosts.

Herd the guests out to the middle of some dark spot and point them in the right direction. Lo and behold: a star as bright as Venus will rise above the horizon and travel across the nocturnal sky in a matter of minutes. Savour the awestruck exclamations of "Wow!" and "Gosh!" and "I think I can see the toilet!"

By the way, it will help your credibility if you have also arranged with NASA to clear the sky of clouds.

rockets. Some steering is performed after lift-off by pivoting the engine nozzles, but the targeting accuracy is comparatively poor and the collateral damage great. The same is true for a person who goes ballistic. The next technological step is hybrid missiles that fly across half the globe in a couple of hours and are guided to their targets by some crew-cut jock in Houston using his laptop and joystick. Gone are the days of innocence when joystick targeting was confined to the privacy of one's bedroom.

THE MOMENT OF TRUTH

The time will come when someone asks where you've acquired your inside knowledge of rocket science. This is when your bluffing skills are put to the ultimate test. Half-hearted answers like "I've read a lot of books on the subject" won't cut it because, with your formidable experience and intellect, you are supposed to be writing books, not reading them. Claiming to have spent 30 years with NASA could get you into trouble with people who have actually done so – and there are quite a few of those.

A safer bet would be to say that you have been doing consultancy work for many years, even if the only important thing you have ever consulted is your watch. It probably won't be lost on your listener that consultants know a lot more than their customers, which is precisely why they are hired. For good measure, you might add: "I have been asked to apply to ...", or "I'm being headhunted for a job with ...", neither of which commits you to anything. But the easiest bluff of all is to look discreetly over your shoulder and then reply, sotto voce: "Between you and me, I'm not allowed to say. It's classified."

"When someone asks where you've acquired your inside knowledge of rocket science... say that you have been doing consultancy work."

THE CROWNING EXPERIENCE

After months of bluffing in the worthy cause of rocket science, you may be ready to take the floor at space conferences, of which there are many, and at various levels of scientific merit. You will be in excellent company, because most speakers at conferences are inveterate bluffers, especially those who claim that their bankrupt space venture is ready for business.

The most irritating bluffers are also the most sought-after speakers. As a purported rocket scientist, you should adopt the irritating habit of mixing your space lingo with marketing babble. Wax and weave about your 'solutions' to non-existent problems. Tell your audience about your institute's plans to 'leverage' its know-how in the marketplace. Use the word 'metrics' regardless of the context, and don't forget to mention 'business model' and 'enabling technologies'. Before you know it, you will be invited to give keynote speeches and chair discussion panels.

As you make your way through the conference circuit, ensure that people know what you stand for. A safe theme to adopt is that, as rocket science takes Man ever further out into the Universe, Planet Earth mustn't be forgotten. Give examples of how satellites help monitor the growth of the hole in the ozone layer and the advance of global warming. Show heart-rending satellite pictures of rain forests going up in smoke. And then drop a bombshell by announcing that, as you speak, our planet is running out of gravity.

“Open with the statement that it doesn't take a rocket scientist to figure out that rockets waste gravity.”

Savour the ensuing silence before you give it all away. Open with the statement that it doesn't take a rocket scientist to figure out that rockets waste gravity. A satellite that weighs a couple of tons on the ground is suddenly weightless in orbit, so the gravity that engendered its weight has obviously been lost. If rockets keep lifting off at the current rate, there will soon be no gravity left. Without gravity, birds will have to fly upside down to stay on Earth. Kangaroos will take their last leap of faith, and bra-makers will go bust. Leave your audience to ponder the following question: Is this the kind of planet we want our grandchildren to inherit? Then depart before the eggs and the tomatoes start flying.

GLOSSARY

Big Bang The defining moment in the creation of the Universe; also the sound a rocket makes when exploding on the launch pad.

Birds Slang for satellites.

Conurbation Not a corn on a foot, but a thorn in the side of rocketry since it curtails overflight opportunities and restricts the allowable directions of launch.

Count-down A habit among rocket scientists to count backward when the end is near.

Gravity gradient The slowly increasing gravitational pull that drags geostationary satellites out of their allocated orbital slots, and makes middle-aged people sign up for Weight Watchers.

Hold An interruption in the count-down when a space rocket has a sudden change of heart.

Jettison A form of abrupt separation implying rejection, like getting rid of excess rocket mass or breaking up with a boyfriend.

Launch vehicle Space babble for 'rocket'.

Luni-solar perturbations Gravitational pull of the Moon and the Sun that threatens to pull satellites out of their orbits. Also a kind of mental instability among rocket scientists.

Remote sensing What some satellites do to map what grows on Earth, and what teenage boys and girls do at parties while the lights are out.

Rocket A member of the salad family with spikey leaves, thought to be an aphrodisiac in Roman times.

Rocket science A scientific approach to growing rocket(s).

Retro-rocket The nearest rocketry equivalent of a brake.

Shock diamonds Precious stones mounted in engagement rings sprung on unsuspecting maidens; also the bead-like structure of a supersonic rocket exhaust plume.

Spacecraft engineering Trying to turn rocket science into something useful.

Sputnik The very first satellite, placed in Earth orbit by the Soviets in 1957. By analogy, the international press variously named the less successful American runner-up Dudnik, Goofnik, Flopnik, Oopsnik and Kaputnik.

Strap-on booster A kind of Viagra for rockets.

Weightlessness A perfect balance of forces, and a middle-aged person's dream.

THE AUTHOR

Peter Berlin blames his parents for his circuitous career path. Their goal was for him to become an astronomer or a nuclear scientist, even though he really wanted to be a fireman.

Having looked in vain for courses in nuclear astronomy, he became a rocket scientist instead. After 25 years with various space agencies, he felt the time was ripe to climb back on the learning curve. To this end, he spends a third of the year accepting consultancy assignments, another third teaching rocket science at universities, and the third third writing his blockbuster novel.

When he returned to one of his former employers looking for consultancy jobs, the latter deported him first to Moscow, then to Kazakhstan, and finally to a remote corner of Siberia. Like Korolev, he eventually re-emerged from the permafrost and went so far as to find a permanent dwelling in the south of England – with occasional lapses as a guest lecturer in darkest Lapland.

Extracts from other Bluffer's® Guides

Genetics

Of course, someone can be a homozygote for one gene (e.g. second toe length), and a heterozygote for another gene (e.g. earwax) all at the same time and still be able to walk and chew gum too.

Consultancy

Always be hard to get. A blank diary must be made to seem full. Bogus meetings must be cancelled or postponed. You must always appear to have had to tear yourself away from urgent and important matters to attend to your client's needs.

The Quantum Universe

Einstein, whose work with light and electrons had opened the curtains on the whole quantomime, wavered between calling quantum mechanics 'incomplete' and declaring its ideas to be 'the system of delusions of an exceedingly intelligent paranoiac, concocted of incoherent elements of thought'.

Public Speaking

One little-known pearl of obscure or irrelevant fact will have more impact and do your reputation more good than any amount of sensible information. Indeed, by delivering it, the speaker is presumed by listeners to know about the subject in depth.

Skiing

There is no such thing as a comfortable ski boot; just concentrate on finding one which doesn't make you pray for an early death. The admission of ski boot agony suggests the probability of inexperience. Experts never whinge about their boots.

The Flight Deck

Psychologists put pilots' stress levels at the top of the league alongside surgeons, although the latter have an advantage – if the scalpel slips they do not accompany the patient to the mortuary. The pilot's stress is rarely apparent, except when the airline loses his suitcase.

The Quantum Universe:

"This book is indispensable. Apart from giving the reader an overview, it is filled to the brim with hilarious anecdotes about the many colourful characters who created quantum mechanics and struggled in vain to make sense of it."

Reader from Belgium

The Classics:

"Packed with a rich store of choice titbits about the ancients, this gives bluffers enough background knowledge to discuss the Classics in an impressively erudite way."

Reader from London

Management:

Fresh, concise and useful this guide offers very reasonable advice to those who aspire to become good managers, and explains fancy terms to the happy bluffer who has no time to read thick books on the subject.

Reader from Switzerland

Archaeology:

"Hilarious truths that archaeologists have tried hard to keep hidden from the public for many years. Having been involved in this bizarre pastime for some years myself, it made me howl with recognition. Buy it!"

Reader from Sheffield

Order form

the Bluffer's® Guides

Accountancy	....	The Olympics	....
Archaeology	....	Opera	....
Astrology	....	Paris	....
Banking	....	Philosophy*	....
Bond*	....	Psychology*	....
The Classics	....	Public Speaking*	....
Consultancy	....	The Quantum Universe*	....
Cricket	....	Relationships	....
Doctors	....	Rocket Science*	....
Economics	....	Rugby	....
The Flight Deck	....	Seduction	....
Football	....	Sex	....
Genetics	....	The Simpsons	....
Golf	....	Skiing	....
Hiking	....	Small Business	....
Jazz	....	Stocks & Shares	....
Life Coaching*	....	Surfing	....
Management*	....	Teaching	....
Marketing	....	University	....
Men	....	Whisky	....
Middle Age	....	Wine	....
Music	....	Women	....
Negotiation*	....		

Oval Books

*This Bluffer's® Guide is available as a downloadable audiobook: **www.audible.co.uk/bluffers**

We like to hear from our readers. Please send us your views on our books and we will publish them as appropriate on our web site: ovalbooks.com.

Oval Books also publish the best-selling Xenophobe's® Guide series – see www.ovalbooks.com

Both series can be bought via Amazon or directly from us, Oval Books through our web site www.ovalbooks.com or by contacting us.

Oval Books charges the full cover price for its books (because they're worth it) and £2.00 for postage and packing on the first book. Buy a second book or more and postage and packing will be entirely FREE.

To order by post please fill out the accompanying order form and send to:

Oval Books
5 St John's Buildings
Canterbury Crescent
London SW9 7QH

cheques should be made payable to: Oval Books

or phone us on +44 (0)20 7733 8585

Payment may be made by Visa or Mastercard and orders are dispatched as soon as the card details and mailing address are received. If the mailing address is not the same as the card holder's address it is necessary to give both.

5 St John's Buildings Canterbury Crescent London SW9 7QH